Dear Reader

UNDER THREAT

Martin Jenkins illustrated by *Tom Frost*

AT THE BEGINNING of the nineteenth century, the passenger pigeon was by far the most common bird in North America and almost certainly in the world, numbering in the billions. In September 1914, Martha, the very last one, died at the Cincinnati Zoo. At the time, it seemed unbelievable that an animal that was once so abundant could disappear so quickly, but it happened.

All the animals in this book are in danger of meeting the same fate as the passenger pigeon. Some are still numerous—though nowhere near as common as the passenger pigeon was—but are declining fast. Others' entire populations are just a handful of individuals. Each species has its own story, and each faces its own particular set of threats, almost always the result of human behavior. But just as humans have created these problems, we are also capable of finding solutions. In some cases—thanks to the tireless efforts of conservationists—we are already doing so. Some of the animals here are in a better state than they were twenty or thirty years ago. They are not out of danger, but their prospects are brighter than they were.

There are huge numbers of threatened species, and the task of saving them all is immense. Perhaps it's all too much effort. Can't we get by without the blue whale or the kakapo or the variable harlequin frog? In truth, we probably can, but I think the world would be a much poorer place without them. And if we go on losing species at the rate we have been, who knows what the effects will be? What kind of world are we going to end up with then?

The stamps in this album aren't real, but the animals and the issues they face all are. They show the extent of the problem and the range of solutions we'll need to invoke to solve it.

Martin Jenkins

CANDLEWICK STUDIO

an imprint of Candlewick Press

CONTENTS

BLUE WHALE

34

ASIAN ELEPHANT

36

IBERIAN LYNX

38

RUSSIAN STURGEON

40

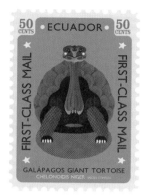

GALÁPAGOS GIANT TORTOISE

42

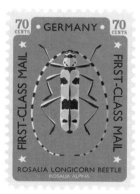

ROSALIA LONGICORN BEETLE

44

GOLDEN LION TAMARIN

46

LARGETOOTH SAWFISH

48

EASTERN GORILLA

50

AFRICAN WILD DOG

52

SUNDA PANGOLIN

54

MEDITERRANEAN MONK SEAL

56

KOREAN CLUBTAIL DRAGONFLY

58

KAKAPO

60

OKAPI

62

RED-CROWNED CRANE

One of the world's largest flying birds, the Asian red-crowned crane has long been revered in Japan as a symbol of longevity and faithfulness. For centuries, cranes were strictly protected in the country, with harsh penalties for anyone caught harming one.

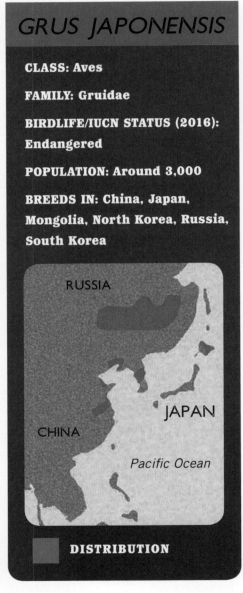

GRUS JAPONENSIS

CLASS: Aves

FAMILY: Gruidae

BIRDLIFE/IUCN STATUS (2016): Endangered

POPULATION: Around 3,000

BREEDS IN: China, Japan, Mongolia, North Korea, Russia, South Korea

RUSSIA

CHINA

JAPAN

Pacific Ocean

■ **DISTRIBUTION**

IN THE NINETEENTH CENTURY, social change swept through Japan and many traditional laws broke down. During these times of hardship, red-crowned cranes became a tempting target for hungry people. The crane population plummeted until it was thought that they had been wiped out in Japan and survived only in mainland northern Asia. In the late 1910s, miraculously, a handful of cranes were discovered in remote marshes in Hokkaido, the northernmost large island of Japan.

At the beginning of the twentieth century, the crane was thought to be extinct in Japan.

The red-crowned crane population clung on, but only in tiny numbers, until the early 1950s, when a program was begun to feed the birds in winter, their most vulnerable time of year. The program was a remarkable success, and numbers have grown steadily since then. As of 2015, there were around 1,500 red-crowned cranes in Hokkaido, about half the total world population. Numbers there are not likely to increase much beyond this, as cranes need big areas of undisturbed swampland to breed in, and there is simply no more room for them. Meanwhile, the population elsewhere in Asia is decreasing as swamps and marshes are drained and roads are built through the cranes' habitat or birds are shot, accidentally poisoned, or killed through collision with power cables.

90 YEN

JAPAN

90 YEN

FIRST-CLASS MAIL

FIRST-CLASS MAIL

RED-CROWNED CRANE

GRUS JAPONENSIS

POLAR BEAR

With their thick white coats, which serve as both camouflage and insulation, and their tremendous strength and keen senses, polar bears are beautifully adapted to a hunting life in the icy Arctic.

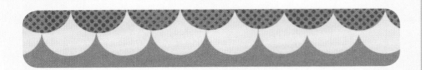

SOMEWHERE BETWEEN 20,000 and 30,000 polar bears live in the Arctic, around two-thirds of them in Canadian territory. Polar bears feed mainly on ringed seals and bearded seals, which they hunt on floating sea ice, either waiting to ambush seals at holes when they come up from the water below to breathe or stalking them when they spend time on the ice to molt or look after their pups. In places where sea ice is always present, polar bears hunt seals year-round. Where the ice melts in the summer, seals move into the open sea, out of the polar bears' reach. Polar bears from these areas may then go for weeks or even months eating little or nothing. Some move inland and hunt reindeer, ground-nesting birds, and other wildlife, forage for grass and berries, or even raid town garbage dumps. But these food sources are not enough to sustain them properly, and they invariably lose weight and become weak.

Until recently, this was not a big problem for the polar bear population overall; much of the Arctic Ocean remained frozen year-round, and in places where the ice did melt, the ice-free period was quite short. With climate change, the ice is melting earlier and freezing later in the year than it used to, and more and more of the Arctic is becoming ice-free in the summer. Preventing this will depend not just on action in the Arctic but on humans worldwide limiting their impact on the global climate. Scientists see halting or reversing climate change as an urgent matter around the world.

At some point, climate change will start to have a drastic effect on polar bear populations.

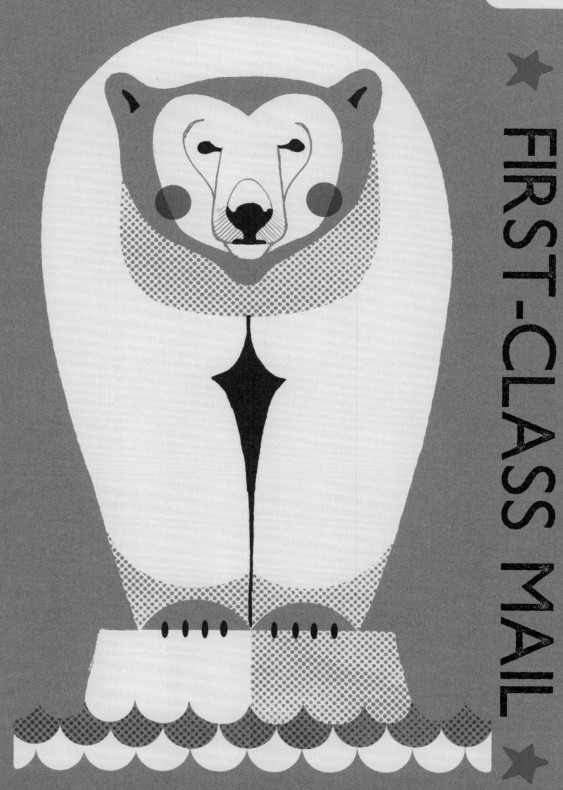

85 CENTS

85 CENTS

CANADA

FIRST-CLASS MAIL

FIRST-CLASS MAIL

POLAR BEAR

URSUS MARITIMUS

GRÉVY'S ZEBRA

Grévy's zebras are elegant long-legged animals, the largest and rarest of the three living species of zebra. They live in drylands in northern and central Kenya and Ethiopia, where they feed on grasses and other low-growing plants.

MOST ADULT MALES LIVE in territories of a few square miles/kilometers that they defend against other males, but females and young males wander over huge areas in search of good grazing, sometimes covering up to 50 miles/80 kilometers in a day. Although well suited to life in a dry climate, Grévy's zebras need to be able to drink regularly—at least once every five days or so, and considerably more often in the case of mothers with young.

By 2008, the population had shrunk to between 2,500 and 3,000.

Up until the late 1970s, Grévy's zebras existed in reasonable numbers, with around 14,000 in Kenya and between 1,500 and 2,000 in Ethiopia. In the early 1980s, however, the population started to collapse—largely, it's believed, because of hunting for the zebra's beautiful coat. By 1990, there were estimated to be only around 5,000 Grévy's zebras left. Increased protection at that time meant that hunting largely stopped, but the population went on declining. This is thought to be mainly because the huge numbers of cattle in the area were preventing the zebras, and particularly mothers with young, from getting to water and finding enough food.

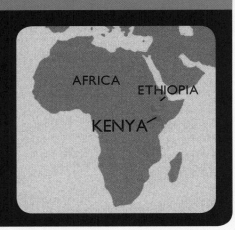

EQUUS GREVYI

CLASS: Mammalia

FAMILY: Equidae

IUCN STATUS (2016): Endangered

POPULATION: Around 2,700

FOUND IN: Ethiopia, Kenya

AFRICA

ETHIOPIA

KENYA

Since 2008, conservationists have stepped up their efforts for the species, including encouraging local cattle owners to share access to water and grazing lands with the zebras, and those efforts seem to be working. The latest figures show that the Grévy's zebra population has stopped declining and may even be slowly increasing.

35 CENTS   KENYA   **35 CENTS**

FIRST-CLASS MAIL

FIRST-CLASS MAIL

GRÉVY'S ZEBRA

EQUUS GREVYI

LAU BANDED IGUANA

The iguanas of the Fiji islands, in the southwest Pacific, are an isolated offshoot of a large family of lizards whose other members are found thousands of miles away, mostly in the Americas, with a few species in Madagascar.

THERE ARE THOUGHT to be four species of iguanas living in various parts of the Fijian archipelago, one of which is the Lau banded iguana. When humans first arrived in this part of the Pacific about 3,000 years ago, there were at least two other, much larger kinds of iguanas, one in Fiji and the other in nearby Tonga. The early settlers hunted them for food and soon drove them to extinction. No one hunts the surviving species much these days, but they face a variety of other threats and are all regarded as endangered, some critically so.

The Lau banded iguana is found in the Lau Islands, a group of about 60 small, scattered islands and islets to the southeast of the main Fijian archipelago. At one time, the iguanas were probably present on most of these islands, but recent surveys have found them on only a few, and in small numbers. The species also occurs on some islands in Tonga, where it was probably introduced by humans around 800 years ago and where it now, too, seems to be rare.

The Lau banded iguana survives on only around a dozen small islands.

The main threats facing the iguanas are predation by rats and by cats, some of which were introduced in an attempt to control rats, as well as loss of the iguanas' forest habitat, often burned down to create pasture for goats. In some places, native forest has been replaced by pine plantations, which are unsuitable for the iguanas. None of the islands where the species occurs is currently protected, so populations will probably continue to decline in the foreseeable future.

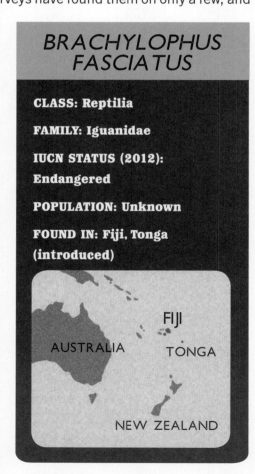

BRACHYLOPHUS FASCIATUS

CLASS: Reptilia

FAMILY: Iguanidae

IUCN STATUS (2012): Endangered

POPULATION: Unknown

FOUND IN: Fiji, Tonga (introduced)

40 CENTS

40 CENTS

FIJI

FIRST-CLASS MAIL

FIRST-CLASS MAIL

LAU BANDED IGUANA

BRACHYLOPHUS FASCIATUS

VARIABLE HARLEQUIN FROG

The tropical rain forests of Central and South America are home to more kinds of plants and animals than anywhere else on Earth. Among them are the harlequin frogs or toads, small, usually brightly colored amphibians that are mostly found along fast-flowing forest streams.

AS WITH MANY RAIN-FOREST SPECIES, many harlequin frogs, most of which live in relatively small areas, are threatened by the loss of their habitat as forests are cleared for farming or cattle ranching and streams are polluted by mining and other human activities. Unfortunately, harlequin frogs face an additional deadly menace: a disease called chytridiomycosis, caused by a fungus. This ailment has devastated amphibian populations worldwide over the past forty years, driving species to extinction or near extinction. Harlequin frogs have been hit harder by this disease than any other amphibian. Of around 100 known species, thirty have not been found in the wild in the past ten years and are feared extinct, and almost all of the rest have seen their numbers plummet.

 **Variable harlequin frog populations began to crash in the late 1980s.**

One of the most striking of the surviving species is the variable harlequin frog from Costa Rica and Panama. Once quite common, by the mid-1990s the species was feared extinct in Costa Rica. At that time, it still survived in good numbers in Panama, but in the early 2000s, it started to decline there too. By good fortune, in 2003 a population was discovered in a small area of protected forest in Costa Rica. It's hoped that, having survived the initial epidemic, this population will persist; some frogs seem to be quite resistant to chytridiomycosis, and these particular harlequin frogs may have developed some immunity. Meanwhile, the species is being bred in captivity in the hope that in the future it may be possible to reintroduce the variable harlequin frog to other areas in the wild where it once lived.

ATELOPUS VARIUS

CLASS: Amphibia

FAMILY: Bufonidae

IUCN STATUS (2008): Critically endangered

POPULATION: Unknown

FOUND IN: Costa Rica, Panama

CENTRAL AMERICA

PANAMA

COSTA RICA

SOUTH AMERICA

450
COLONES

450
COLONES

COSTA RICA

FIRST-CLASS MAIL

FIRST-CLASS MAIL

VARIABLE
HARLEQUIN FROG
ATELOPUS VARIUS

KEMP'S RIDLEY

There are seven kinds of sea turtles in the world today. Kemp's ridley, whose home is the Gulf of Mexico, is the smallest of these, and by far the rarest.

LEPIDOCHELYS KEMPII

CLASS: Reptilia

FAMILY: Cheloniidae

IUCN STATUS (1996): Critically endangered

POPULATION: Unknown

BREEDS IN: Mexico, United States

LIKE ALL SEA TURTLES, Kemp's ridleys spend essentially their whole life at sea, with the females coming ashore only to lay their eggs in the nesting season. The species was known to breed along the US and Mexican coasts of the Gulf of Mexico, but for many years the nesting site of the main population was a mystery. Then, in the 1960s, a 1947 film clip was unearthed showing around 40,000 of the turtles emerging all at once onto a beach in the town of Rancho Nuevo in Tamaulipas, Mexico. But when the area was surveyed in the mid-1960s, it became clear that this nesting population—by far the largest in the world—had collapsed. For years people had been gathering almost all the eggs, and sometimes the turtles themselves, for food. In addition, many turtles were dying at sea after being trapped in shrimp nets or caught on fishing lines.

Despite legal protection of the turtles and their nesting beach, the situation deteriorated. Conservation efforts at Rancho Nuevo were stepped up in the 1980s: fishing regulations were tightened, and many nets were equipped with Turtle Excluder Devices (TEDs), which allow trapped turtles to escape. Also, eggs were transported to Padre Island in Texas to bolster the tiny US breeding population. At first, numbers slowly began to increase. In 2009, there were around 20,000 nests (each female breeds two to three times a year), including around 200 in the United States, almost all at Padre Island. Since then, however, the total number of nests each year has gone down somewhat. Scientists do not know why, but they are working hard to solve the problem.

By the mid-1980s, fewer than 300 turtles were nesting each year.

10 PESOS · MEXICO · 10 PESOS

FIRST-CLASS MAIL

FIRST-CLASS MAIL

KEMP'S RIDLEY

LEPIDOCHELYS KEMPII

INDRI

Madagascar is a unique place. The world's fourth-largest island, it is home to thousands of kinds of animals and plants that are found nowhere else in the world. Among them are lemurs, which are primates related to monkeys, apes, and humans.

WHEN THE FIRST HUMANS arrived on Madagascar, probably between 2,000 and 3,000 years ago, there were lemurs there the size of calves. These giants are long gone, very likely quickly hunted to extinction, but around 100 smaller kinds still survive, the largest of which is the indri, or babakoto, as it is known locally. Indris live in small family groups in rain forests on the eastern side of the island. They spend most of their time high up in the trees, feeding on tender young leaves and periodically filling the air with their extraordinary wailing songs.

According to local legend, indris are descended from humans that strayed into the forest.

Due to traditional beliefs about the sacredness of the indri, hunting them has often been banned by local custom. Like all lemurs, they are also in theory protected by official laws. However, these laws are generally not well enforced, and in some places, traditional protection is falling away too, so hunting now poses a real problem to the indri population. Even more critically, the area of rain forest in eastern Madagascar grows smaller each year as land is cleared for crops. Even the forests in national parks and nature reserves are not safe. Madagascar is one of the world's poorest countries, and it simply doesn't have the resources to protect these areas effectively on its own. Help from overseas has gone some way to filling the gap, but more is undoubtedly needed. Without it, the future of the indri and much of the country's other unique and extraordinary wildlife looks bleak indeed.

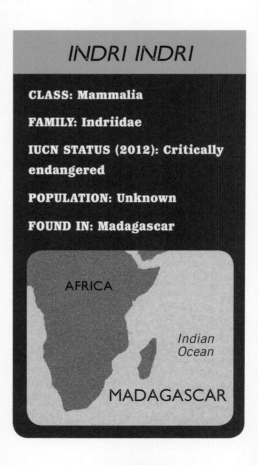

INDRI INDRI

CLASS: Mammalia

FAMILY: Indriidae

IUCN STATUS (2012): Critically endangered

POPULATION: Unknown

FOUND IN: Madagascar

AFRICA

Indian Ocean

MADAGASCAR

3000 ARIARY

3000 ARIARY

MADAGASCAR

FIRST-CLASS MAIL

FIRST-CLASS MAIL

INDRI

INDRI INDRI

TIGER

Not so long ago, tigers, the largest of all cats, were widespread in Asia from the Caspian Sea to the Far East. All that has changed. They have disappeared from much of this area, and the populations that do survive are small and scattered. It's not difficult to understand why.

HUMANS AND TIGERS do not mix very well, and many places where the tiger used to be found now have huge numbers of people living there. Tigers are top predators that need a lot of room and a plentiful supply of prey, such as deer, wild pigs, and antelopes. Where these animals have become rare, tigers are likely to turn instead to domestic animals, especially sheep, cattle, and goats, and sometimes even people. What's more, tiger pelts are highly sought-after as trophies, and their bones are a valuable ingredient in traditional Chinese medicine. All this makes tigers a target for poachers.

The incentive to hunt tigers is strong, even though it is against the law everywhere.

Where protection is effective, however, and there is enough space and a good number of prey animals, tigers can flourish. Even in India, one of the most densely populated countries in the world, there are some places where tigers are doing well. India is home to around two-thirds of all wild tigers, and numbers there may actually have increased in recent years—in 2011, there were believed to be around 1,700 in India, while a census in 2014 concluded there were more than 2,200. Sustaining these numbers in the face of an ever-growing human population is a great challenge, but one that is undoubtedly achievable.

PANTHERA TIGRIS

CLASS: Mammalia

FAMILY: Felidae

IUCN STATUS (2014): Endangered

POPULATION: 3,000–4,000

FOUND IN: Bangladesh, Bhutan, China, India, Indonesia, Laos, Malaysia, Myanmar, Nepal, Russia, Thailand. May still survive in Cambodia, North Korea, Vietnam.

RUSSIA

INDIA

5 RUPEES

5 RUPEES

INDIA

FIRST-CLASS MAIL

FIRST-CLASS MAIL

TIGER

PANTHERA TIGRIS

GIANT PANDA

Among the most instantly recognizable of all wild animals and one of the most well-known threatened species, the giant panda is a highly unusual vegetarian member of the bear family.

GIANT PANDAS FEED almost entirely on bamboo. They were once widespread in bamboo forests in northern parts of Southeast Asia and in China as far north as Beijing. Over the years, their range has shrunk through a combination of hunting and habitat loss as forests have been cleared for agriculture. By the middle of the twentieth century, giant pandas were confined to mountainous areas in Sichuan, Shaanxi, and Gansu provinces in western China. A survey in the 1970s found them still reasonably common there, with a total population of around 2,500. However, a second survey in the 1980s uncovered alarming evidence of a rapid decline—linked to illegal hunting and ongoing forest loss. Researchers counted only half as many giant pandas as they had ten years before.

There are now nearly seventy giant panda reserves in western China, protecting a total of around 3.5 million acres/1.4 million hectares of habitat.

In response to the decline, a major conservation program was launched, involving the establishment of a network of reserves and a redoubling of efforts to stamp out hunting. Following the introduction of a national reforestation policy in China in 1997, the area of forest within the giant panda's range has also increased. The most recent survey, between 2011 and 2014, found around 2,000 wild giant pandas in total, a marked increase since the 1980s.

Giant pandas are still not entirely secure. Many of the surviving populations are small and isolated, making them vulnerable to dying out through accidents or disease, and climate change is projected to start causing a decrease in the amount of suitable habitat in the coming decades. Nevertheless, the species is in a far better position now than it was thirty years ago.

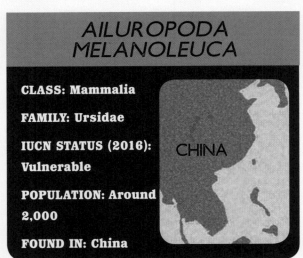

AILUROPODA MELANOLEUCA

CLASS: Mammalia

FAMILY: Ursidae

IUCN STATUS (2016): Vulnerable

POPULATION: Around 2,000

FOUND IN: China

CHINA

1.20 YUAN

CHINA

1.20 YUAN

FIRST-CLASS MAIL

FIRST-CLASS MAIL

GIANT PANDA

AILUROPODA MELANOLEUCA

NUMBAT

Australia is home to a wide range of extraordinary animals found nowhere else. These include the numbat, also known as the marsupial anteater — although it feeds on termites, not ants.

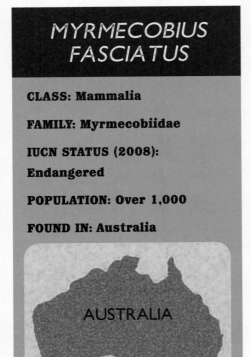

MYRMECOBIUS FASCIATUS

CLASS: Mammalia

FAMILY: Myrmecobiidae

IUCN STATUS (2008): Endangered

POPULATION: Over 1,000

FOUND IN: Australia

AUSTRALIA

A S GENERATIONS of Europeans arrived on the continent, beginning in the late 1700s, they brought with them, deliberately or accidentally, a host of other animals and plants, many of which ended up having a huge impact on native species. Among the most disastrous of those introductions was the red fox, brought over in the nineteenth century by people wishing to indulge in fox hunting, a popular activity in Britain at that time. After several failed attempts, a population of foxes became established in the 1870s, and it quickly spread across the southern half of the country. The effect on many animals, completely unused to this new predator, was devastating. Some native species became entirely extinct, and others survived only on offshore islands. In each case the fox is thought to have been the main culprit, although other factors such as habitat destruction and predation by feral cats may also have played a part.

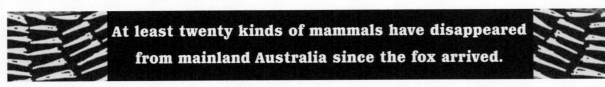

At least twenty kinds of mammals have disappeared from mainland Australia since the fox arrived.

Numbats have not been completely wiped out, and they still retain a toehold on the Australian mainland. They were once widespread in open woodland across the whole of southern Australia, but now natural populations of this bushy-tailed, pointy-nosed mammal survive only in two fox-free nature reserves in southwest Australia. There are also a few reintroduced numbat populations in Western Australia and in two sanctuaries, one in New South Wales and one in South Australia, that are fenced to keep foxes and feral cats out.

$1 $1

AUSTRALIA

FIRST-CLASS MAIL

FIRST-CLASS MAIL

NUMBAT

MYRMECOBIUS FASCIATUS

CALIFORNIA CONDOR

With a 10-foot/3-meter wingspan and weighing up to 31 pounds/14 kilograms, the California condor is North America's largest land bird. It was once a widespread scavenger, but its range began to shrink thousands of years ago, when humans arrived in North America.

B Y THE NINETEENTH CENTURY, the condor had become confined to the west coast of the United States. Here its decline continued, due mainly to ranchers who falsely believed it was killing their livestock, and to lead poisoning from ingesting ammunition lodged in the carcasses on which the condors fed.

By 1981, there were just twenty-two California condors left in the wild, all in Southern California. The situation was desperate. Wildlife officials decided that their only option was to take the remaining birds into captivity in the hope of breeding them and building up a population to release back into the wild. It was a controversial and risky undertaking, but it worked. In 1988, the first captive-bred chick hatched, and in 1992, two condors raised in captivity were released in California. More releases followed in Arizona as well as California and, in 2002, in the Mexican state of Baja California.

 **Condors in the wild finally started breeding again in 2002.**

Numbers have gradually increased, but it's a slow process—condors do not breed until they are six years old, and they raise only one chick every other year at most. By the end of 2016, there were 276 California condors in the wild and 170 in captivity. The wild population continues to depend on the release of birds from captivity to sustain itself. With luck, that will eventually change. Following a law passed in California in 2008, nearly all hunters there now use lead-free ammunition, but elsewhere lead poisoning is still a major threat to condors.

GYMNOGYPS CALIFORNIANUS

CLASS: Aves

FAMILY: Cathartidae

BIRDLIFE/IUCN STATUS (2017): Critically endangered

POPULATION: 446 (276 in the wild, 170 in captivity)

FOUND IN: Mexico, United States

UNITED STATES

MEXICO

Pacific Ocean

50 CENTS

50 CENTS

UNITED STATES of AMERICA

FIRST-CLASS MAIL

FIRST-CLASS MAIL

CALIFORNIA CONDOR

GYMNOGYPS CALIFORNIANUS

BLACK RHINOCEROS

Until fifty years ago, the black rhinoceros was by far the most common of the five living rhino species. A solitary, famously pugnacious creature, it was found in a range of habitats across much of Africa south of the Sahara and had a population estimated at 100,000 or more.

SOON, THOUGH, IT FELL into a precipitous decline, caused by a huge increase in hunting for its horn. During the 1970s and 1980s, rhino horns were used in making handles for *jambiya* daggers in Yemen and in traditional Chinese medicine, in which it was prescribed to treat a variety of ailments, especially fever, rheumatism, and gout. The market for rhino horn peaked in Yemen in the 1980s, but demand in China and other Asian countries continued to grow, fueled by the economic growth in the region. Commercial international trade in all rhino products has been banned under the Convention on International Trade in Endangered Species (CITES) since 1977, and all rhinos are legally protected. But poaching continues.

The high value of rhino horn has been a powerful incentive for poachers.

By 1995, the black rhino population had fallen to fewer than 2,500. The only places where numbers were increasing were South Africa and Namibia, two countries that had invested heavily in wildlife protection. Since then, populations have stabilized or increased in a number of other countries, as they too have stepped up protection measures. By the end of 2015, there were estimated to be 5,250 black rhinos, with most in South Africa and Namibia (around 1,900 each), and almost all the remainder in Kenya, Tanzania, and Zimbabwe.

Conserving rhinos in the face of illegal killing and trade is an ongoing battle. At the moment, the main demand comes from Vietnam, where rhino horn is seen as a prestige item by some successful business executives. Trying to change this attitude is as important a part of conservation efforts as protecting rhinos on the ground.

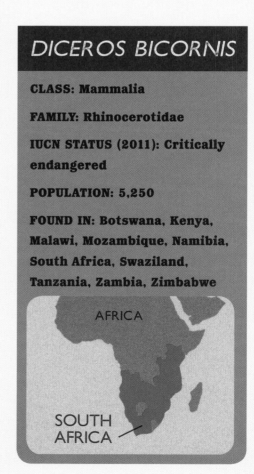

DICEROS BICORNIS

CLASS: Mammalia

FAMILY: Rhinocerotidae

IUCN STATUS (2011): Critically endangered

POPULATION: 5,250

FOUND IN: Botswana, Kenya, Malawi, Mozambique, Namibia, South Africa, Swaziland, Tanzania, Zambia, Zimbabwe

AFRICA

SOUTH AFRICA

R
6,35

R
6,35

SOUTH
AFRICA

FIRST-CLASS MAIL

FIRST-CLASS MAIL

BLACK RHINOCEROS

DICEROS BICORNIS

EUROPEAN EEL

ANGUILLA ANGUILLA

CLASS: Actinopterygii

FAMILY: Anguillidae

IUCN STATUS (2013): Critically endangered

POPULATION: Unknown

FOUND (AS ADULTS) IN: 56 countries in western Europe, North Africa, and the Mediterranean

For centuries the life cycle of the European eel was a mystery — no one had ever seen the eels breed or discovered where the young eels lived. Then, in the 1900s, scientists figured out that tiny sea creatures from the Atlantic Ocean called leptocephali were in fact baby eels.

BY MAPPING WHERE the leptocephali were found, scientists deduced that European eels make an incredible 4,300-mile/7,000-kilometer journey across the Atlantic Ocean to the Sargasso Sea, near the United States, in order to breed, and that the resulting leptocephali then gradually make their way back to Europe. Until recently, they did so in huge numbers, and the eels were a common species in fresh waters and around the coasts of western Europe, where they were widely harvested for food—in the United Kingdom especially for jellied eels, a Cockney delicacy.

Since the 1980s, eel stocks have plummeted. Many different factors have been blamed, including pollution, particularly from chemicals known as PCBs (polychlorinated biphenyls), which are thought to prevent eels from reproducing successfully, and changes in ocean currents in the Atlantic, which interfere with eel migration. Overfishing, diseases and parasites (especially a nematode worm accidentally introduced from Asia), and barriers such as dams, which stop eels from migrating up and down rivers, are also thought to have had a negative impact. It is not clear whether just one of these factors is the main cause of the problem or if it is due to several acting in combination.

Eel numbers have dropped in some places to less than one-hundredth of what they were.

Because of the uncertainty, it is hard to know how best to improve things. Use of PCBs is now generally banned in Europe and the United States, but they can persist in the environment for years, so it will be a long time before the ban shows positive results. Fishing regulations have been strengthened, and efforts have been made to allow eels to move up and down some rivers more easily. Restocking has also taken place, but because no one has yet successfully bred European eels in captivity, wild-caught young eels still need to be moved from one place to another.

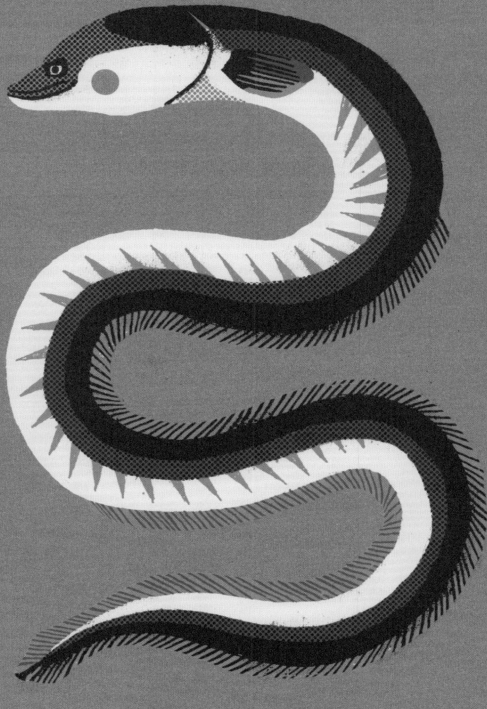

65 PENCE

65 PENCE

UNITED KINGDOM

FIRST-CLASS MAIL

FIRST-CLASS MAIL

EUROPEAN EEL

ANGUILLA ANGUILLA

TAPANULI ORANGUTAN

Orangutans belong to the same family of primates as humans, gorillas, chimpanzees, and bonobos, collectively known as great apes.

APART FROM HUMANS, which are probably the most common large animal the world has ever seen, all great apes share an important characteristic: they are more or less all seriously threatened with extinction.

Until recently, there were generally considered to be two orangutan species living in rain forests in Southeast Asia: one on the island of Borneo and the other in northern Sumatra, an Indonesian island. It now appears that there are not two species, but three. Since the 1930s, rumors had circulated of a population of orangutans in mountainous country in an area known as Batang Toru, south of the known orangutan range in Sumatra. In the late 1990s, scientists confirmed that orangutans did indeed survive here. After careful study, they concluded that these orangutans are different enough from other orangutans to merit being described as a separate species: the Tapanuli orangutan, with a population estimated at just 800 or so individuals.

This orangutan is by far the rarest of all great ape species.

Efforts to conserve the Tapanuli orangutan have been under way since its discovery, and most of the approximately 600 square miles/1,000 square kilometers of forest within the orangutans' range are now officially protected. No legal logging has taken place there since 2001, but threats still remain: land continues to be cleared for farming in some areas by people moving in from other districts; there is an expanding open-pit gold mine in the region; and a hydroelectric dam is planned, which, if it goes ahead, will flood a significant area of forest. Occasionally, orangutans are killed when they come into conflict with local people because of crop damage, and some illegal trade in young orangutans as pets has been reported.

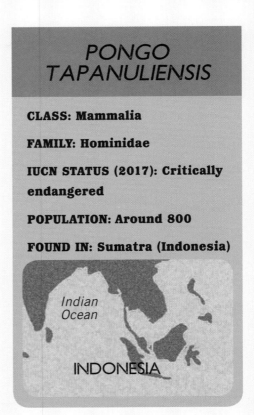

PONGO TAPANULIENSIS

CLASS: Mammalia

FAMILY: Hominidae

IUCN STATUS (2017): Critically endangered

POPULATION: Around 800

FOUND IN: Sumatra (Indonesia)

Indian Ocean

INDONESIA

6000 RP

INDONESIA

6000 RP

FIRST-CLASS MAIL

FIRST-CLASS MAIL

TAPANULI ORANGUTAN
PONGO TAPANULIENSIS

AMSTERDAM ALBATROSS

The entire world population of the Amsterdam albatross, one of the largest flying birds on Earth, breeds in one tiny patch of upland peat bog on Amsterdam Island, a French territory in the southern Indian Ocean.

IN 2014, THE BREEDING population was estimated at forty-six pairs, a tiny number but still a marked improvement from the early 1980s, when there were just five pairs. Not all of the pairs breed every year—couples that have successfully raised a chick one season take a year off at sea before trying again, and young birds spend several years out at sea before returning to try to find a partner of their own.

In the 1980s, the Amsterdam albatross was on the brink of extinction.

Like other albatrosses, the Amsterdam albatross faces a number of threats, including accidental killing in longline fisheries. The albatrosses are attracted to the baited fishhooks that are let out from the fishing boats, but they can get caught on them and drown. There was a lot of tuna fishing around Amsterdam Island in the 1970s and 1980s, and it almost certainly contributed to the bird's extreme rarity at that time. In addition, its breeding site was heavily disturbed by wild cattle, which had been introduced to the island in 1871.

In 1992, the breeding site was fenced off, and in 2011, cattle were eliminated from Amsterdam Island. There has also been less tuna fishing around the island in the past couple of decades, and measures introduced to try to reduce the deaths of seabirds in longline fisheries seem to be paying off. The main threat to the Amsterdam albatross is now thought to be diseases such as avian cholera, which can be fatal, especially to chicks. In 2000 and 2001, over two-thirds of the young birds in the colony died. Luckily this has not happened again since then.

DIOMEDEA AMSTERDAMENSIS

CLASS: Aves

FAMILY: Diomedeidae

BIRDLIFE/IUCN STATUS (2016): Critically endangered

POPULATION: Around 200 (including around 50 breeding pairs)

BREEDS IN: Amsterdam Island (French Southern Territories)

MADAGASCAR

AUSTRALIA

FRENCH SOUTHERN TERRITORIES

95 CENTS

95 CENTS

FIRST-CLASS MAIL

FIRST-CLASS MAIL

AMSTERDAM ALBATROSS

DIOMEDEA AMSTERDAMENSIS

BLUE WHALE

As far as we know, blue whales are the largest animals that have ever lived. They were once common, too, found in all of the world's main oceans.

BLUE WHALES FEED ON SMALL free-swimming crustaceans called krill. The biggest concentrations of krill are found in the Southern Ocean around Antarctica, and that is where the largest numbers of blue whales are found, at least during the summer months, when the whales are feeding. They generally stop eating in the winter and move into warmer waters to breed and give birth.

Being the largest whales, historically blue whales were highly valuable targets for whalers, mainly for their oil. However, their speed and power meant that few were caught until the 1860s, when harpoon cannons started to be used in the North Atlantic. By 1900, blue whales had been hunted to near extinction there, and attention turned to the Southern Ocean, where over the next few decades far more were killed than the population could sustain. In 1946 the International Whaling Commission was set up to try to limit the numbers of all great whales caught, but it proved mostly ineffective, and the numbers of blue whales continued to fall.

In the early twentieth century, many thousands of blue whales were killed in the Southern Ocean each year.

Finally, in 1966, all commercial whaling of the species was outlawed, although some illegal hunting continued until the 1970s. By this time, the Southern Ocean population, thought to have numbered around a quarter of a million in the early 1900s, had been reduced to a few hundred. There were fears that there might be too few left for the species to recover, but fortunately this doesn't seem to have been the case. The population in the Southern Ocean and in other parts of the world is now growing at a reasonable rate, although it is still only a tiny percentage of what it was a century ago.

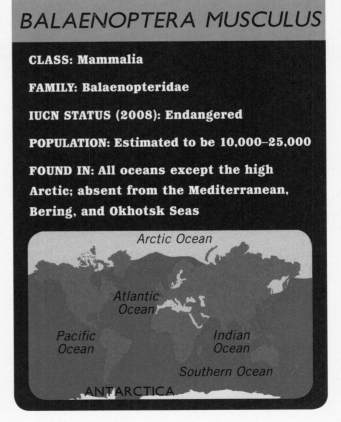

BALAENOPTERA MUSCULUS

CLASS: Mammalia

FAMILY: Balaenopteridae

IUCN STATUS (2008): Endangered

POPULATION: Estimated to be 10,000–25,000

FOUND IN: All oceans except the high Arctic; absent from the Mediterranean, Bering, and Okhotsk Seas

Arctic Ocean

Atlantic Ocean

Pacific Ocean

Indian Ocean

Southern Ocean

ANTARCTICA

66 CENTS

66 CENTS

ANTARCTICA

FIRST-CLASS MAIL

FIRST-CLASS MAIL

BLUE WHALE

BALAENOPTERA MUSCULUS

ASIAN ELEPHANT

There are two kinds of elephants alive today: African and Asian. The Asian elephant is by far the rarer, having a total population of somewhere between 40,000 and 50,000, about one-tenth that of its western cousin.

CENTURIES AGO, ASIAN ELEPHANTS ranged all the way from Iran across the southern part of Asia to the Yangtze, or Chang, River basin in China. A combination of hunting and habitat destruction has led to their disappearance from much of this area, but they are still found in parts of mainland South and Southeast Asia and on the islands of Sumatra and Borneo.

The elephants on Borneo are something of a puzzle. Genetic studies have shown that they are quite distinct from those that survive elsewhere. It is possible that they represent a population that arrived in Borneo naturally hundreds of thousands of years ago, at a time when sea levels were much lower than they are today. Alternatively, they may be the descendants of captive elephants brought to Borneo in the past few hundred years, probably originally from Java, where wild elephants have since become extinct. People have been taming Asian elephants for at least 2,000 years, probably longer, and there is a long tradition of trade in them in the region.

Across the elephant's range, large areas of forest have been cleared for agriculture.

Whatever their origin, the Bornean elephants now represent an important population of this threatened species. There are currently estimated to be around 1,500 here, about the same number as in peninsular Malaysia. Almost all of the elephant's range in Borneo lies within the Malaysian state of Sabah, with a small part in Indonesian Kalimantan. Much of the region is now given over to oil-palm plantations and other forms of agriculture, but fortunately there are also some extensive areas of forest set aside as wildlife reserves. As long as these areas continue to be protected, at least some elephants should survive here.

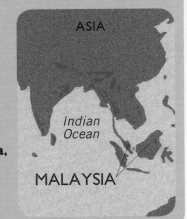

ELEPHAS MAXIMUS

CLASS: Mammalia

FAMILY: Elephantidae

IUCN STATUS (2008): Endangered

POPULATION: 40,000–50,000

FOUND IN: Bangladesh, Bhutan, Cambodia, China, India, Indonesia, Laos, Malaysia, Myanmar, Nepal, Sri Lanka, Sumatra (Indonesia), Thailand, Vietnam

ASIA

Indian Ocean

MALAYSIA

60 SEN · MALAYSIA · **60 SEN**

★ FIRST-CLASS MAIL ★

★ FIRST-CLASS MAIL ★

ASIAN ELEPHANT
ELEPHAS MAXIMUS

IBERIAN LYNX

Once thought to be a form of the widespread Eurasian lynx, the Iberian lynx is now known to be a distinct species. It is one of the world's rarest cats, though its prospects for survival have recently improved considerably.

LIKE MOST WILD CATS, the Iberian lynx is secretive, solitary, and mainly nocturnal. It is a specialized hunter, feeding almost entirely on wild rabbits; its favored habitat is a mixture of Mediterranean scrubland and pasture, where rabbits are abundant and there is good cover for hunting and plenty of suitable sites to make a den.

Up until the 1950s, the Iberian lynx was widespread and quite common in Spain and Portugal, with a population numbering in the thousands. However, in 1953, the disease myxomatosis, which affects rabbits, arrived in Spain. Rapidly spreading through the country, it reduced rabbit populations by as much as 99 percent in some areas, and this in turn caused lynx numbers to collapse. Eventually rabbit populations began to recover as the disease became less virulent, but lynx numbers stayed low. The lynx were still struggling to find enough to eat, as the rabbits were now being intensively hunted by people for food and fur. On top of this, the lynx themselves were targeted by hunters, who feared they were preying on livestock. All of this had a disastrous impact: by 2000, there were believed to be just sixty or so breeding-age Iberian lynx left in the wild.

Concerted efforts to save the species have been undertaken since then. These include working to secure the lynx's food supply—for example, by limiting who can legally hunt rabbits in areas where the lynx live—and ensuring that there are enough denning sites and water holes for them. In addition, the lynx has been reintroduced in a number of places. By 2016, the breeding population had increased to around 240 animals spread across four sites in Spain and one in Portugal. The main threat now is road traffic: more than fifty lynx were killed on roads between 2014 and 2016, representing a serious drain on the population.

LYNX PARDINUS

CLASS: Mammalia

FAMILY: Felidae

IUCN STATUS (2014): Endangered

POPULATION: 400 (including around 240 adults)

FOUND IN: Spain, Portugal (reintroduced)

 **By 2000, virtually all of the remaining lynx were confined to just two areas in southwest Spain.**

55 CENTS

55 CENTS

SPAIN

FIRST-CLASS MAIL

FIRST-CLASS MAIL

IBERIAN LYNX

LYNX PARDINUS

RUSSIAN STURGEON

Sturgeons are a family of large, long-lived fish whose main claim to fame is that they produce that most luxurious and expensive of foodstuffs: caviar.

CAVIAR—THE SALTED EGGS of female sturgeons—comes in various forms. Beluga is the most sought-after, but second to this is osetra caviar, made from the eggs of the Russian or diamond sturgeon, a medium-size species from the basins of the Black and Caspian Seas and the Sea of Azov.

Like the majority of other sturgeons, Russian sturgeons typically spend most of their life at sea, usually in relatively shallow coastal waters, but breed in rivers, often swimming hundreds of miles upstream to suitable spawning sites. There are, or at least there used to be, some entirely freshwater populations, too.

Restocking operations have been undertaken in an attempt to keep the wild sturgeon population going.

Until the middle of the twentieth century, the Russian sturgeon bred in most of the large rivers within its range. Since then, wild populations have declined catastrophically for two main reasons: the building of hydroelectric dams on rivers, which has blocked access to most spawning sites, and unregulated or illegal fishing for caviar, particularly since the collapse of the Soviet Union in 1991. In the past twenty years, spawning has been recorded with certainty only in the lower parts of the Danube River in the Black Sea basin, and in the Ural and Volga Rivers in the Caspian basin.

Millions of young fish have been raised in captivity over the years and released mainly by Russia and Iran into the Caspian Sea to try to increase the wild population. This effort has met with only limited success, and the number of fish spawning in the wild has continued to decline. The species as a whole, though, is unlikely to go extinct in the immediate future, as large numbers are kept in captivity, mainly for the production of caviar.

ACIPENSER GUELDENSTAEDTII

CLASS: Actinopterygii

FAMILY: Acipenseridae

IUCN STATUS (2009): Critically endangered

POPULATION: Unknown

FOUND IN: Azerbaijan, Bulgaria, Georgia, Iran, Kazakhstan, Moldova, Romania, Russia, Serbia, Turkey, Turkmenistan, Ukraine

22 ROUBLES

22 ROUBLES

RUSSIA

FIRST-CLASS MAIL

FIRST-CLASS MAIL

RUSSIAN STURGEON

ACIPENSER GUELDENSTAEDTII

GALÁPAGOS GIANT TORTOISE

In 1535, a Spanish ship traveling from Panama to Peru drifted off course in the Pacific Ocean and came across a group of islands inhabited by a rich variety of wildlife, including sea lions, flightless cormorants, and seagoing lizards.

THE MOST DISTINCTIVE of the animals the crew encountered were giant tortoises, found in the thousands. These were to give their name to the archipelago, called the Galápagos Islands, after the Spanish word for tortoise. At the time of their discovery, the tortoises lived on at least seven of the eighteen main islands. Today they survive on five. Populations on different islands, and in different parts of Isabela, the largest island, vary in appearance. Some scientists regard these groups as separate species, while others consider them all as belonging to the same species.

Sailors stored tortoises on board their ships and used them for fresh meat on long voyages.

CHELONOIDIS NIGER
COMPLEX

CLASS: Reptilia

FAMILY: Testudinidae

IUCN STATUS (2015): Vulnerable (based on assessment of least threatened populations)

POPULATION: 20,000–25,000

FOUND IN: Galápagos Islands (Ecuador)

GALÁPAGOS ISLANDS
ECUADOR

Following their discovery, the Galápagos Islands started to become a stop off for explorers, pirates, and, from the 1790s onward, whaling ships and seal hunters, all seeking provisions. Unfortunately for the tortoises, they proved ideal for this. Later the tortoises were also collected for their oil, used in mainland Ecuador. In the nineteenth century, settlers arrived on the islands and brought with them goats and cattle, which competed with the tortoises for food, and pigs and rats, which dug up tortoise eggs or ate the young. The archipelago was declared a national park in 1959, but little action took place to protect the tortoises until the 1970s, by which time the entire population had shrunk to around 3,000. Since then, conservation efforts have been stepped up, with the introduced rats, pigs, and goats eliminated or effectively controlled in some areas, and captive-bred tortoises released to bolster the wild population. The number of giant tortoises now stands at over 20,000.

50 CENTS

50 CENTS

ECUADOR

FIRST-CLASS MAIL

FIRST-CLASS MAIL

GALÁPAGOS GIANT TORTOISE

CHELONOIDIS NIGER SPECIES COMPLEX

ROSALIA
LONGICORN BEETLE

There are more kinds of beetles than any other type of animal. Most are small and found in tropical rain forests, but there are many exceptions. Among them is the spectacular Rosalia longicorn beetle.

IT IS A LARGE BEETLE, with a body up to two inches/four centimeters long, which lives in woodlands and forests in southern and central Europe and western Asia. Like most insects, it spends most of its life as a larva, in this case two to four years, compared with a few weeks at most as an adult. The larvae feed on the dead and decaying wood of hardwood trees, and in central Europe almost exclusively on beech. Because this wood needs to be reasonably dry, the beetles avoid fallen trees and rotting logs in damp, shady places, instead preferring upright dead trees or damaged branches on living trees in sunny spots.

Modern forestry has greatly diminished the habitat available to these beetles.

Unfortunately, in many areas, hardwoods have been replaced by faster-growing conifers. Where hardwoods such as beech are still grown, dead, damaged, or diseased trees are typically removed, leaving nowhere for the larvae to feed. Even worse, where trees are managed for the production of firewood or timber, it is common for the cut wood to be stacked in open areas or on the edge of woodland to dry out. To female beetles, these woodpiles are ideal sites in which to lay their eggs. Of course, virtually none of the resulting young ever grow to be an adult, as the wood has been burned or turned into furniture long before then. As a result, Rosalia longicorn beetle populations have plummeted in many areas, including southern Germany, where it was once relatively common.

ROSALIA ALPINA

CLASS: Insecta

FAMILY: Cerambycidae

IUCN STATUS (1996): Vulnerable

POPULATION: Unknown

FOUND IN: Albania, Austria, Belarus, Bulgaria, Croatia, Czech Republic, France, Germany, Greece, Hungary, Italy, Liechtenstein, Montenegro, Moldova, Poland, Portugal, Romania, Russia, Serbia, Slovenia, Spain, Switzerland, Turkey, Ukraine

GERMANY

EUROPE

70 CENTS

70 CENTS

GERMANY

FIRST-CLASS MAIL

FIRST-CLASS MAIL

ROSALIA LONGICORN BEETLE

ROSALIA ALPINA

GOLDEN LION TAMARIN

Five hundred years ago, the plains and foothills of southeastern Brazil were covered in dense forests that harbored a rich array of species, many of which were found nowhere else.

CENTURIES OF LAND CLEARANCE for growing sugar and coffee and, most recently, for cattle ranching have reduced these forests to small remnants, driving many of the species that live in them to near extinction. Among these is the golden lion tamarin, a small, richly colored monkey with a luxurious mane. Once common in coastal parts of Rio de Janeiro state, by the 1960s, there were only approximately 600 left, many of them living in tiny patches of forest where they had little hope of surviving in the long term. About half lived in just one protected area, the Poço das Antas Biological Reserve, where there were around 11 square miles/27 square kilometers of forest.

By the 1960s, just a few scattered populations were left.

Conservation efforts started in earnest in the early 1970s. Zoos that had captive golden lion tamarins joined in an organized breeding program with the aim of reintroducing the animals into the wild, and family groups from isolated, unprotected forest patches were translocated to larger, more secure areas. These initiatives stopped the decline, but it was not until the late 1990s that numbers began to build up again. There are now over 3,000 golden lion tamarins in the wild, around a third of which are descended from animals bred in captivity, and another 500 or so in zoos. The main problem now is that there are no more areas of suitable, well-protected forest within the tamarin's former range where new populations can be established.

This might change. Over the past decade, millions of native tree seedlings have been planted in southeastern Brazil, with ambitious plans for many millions more to be planted. This should greatly increase the amount of habitat available for the tamarins.

LEONTOPITHECUS ROSALIA

CLASS: Mammalia

FAMILY: Callitrichidae

IUCN STATUS (2008): Endangered

POPULATION: 3,700 (3,200 in the wild, around 500 in captivity)

FOUND IN: Brazil

SOUTH AMERICA

BRAZIL

1,40 REAIS

1,40 REAIS

BRAZIL

FIRST-CLASS MAIL

FIRST-CLASS MAIL

GOLDEN LION TAMARIN
LEONTOPITHECUS ROSALIA

LARGETOOTH SAWFISH

Sawfish are a highly distinctive — and highly threatened — family of fish related to sharks and rays. They live in shallow waters around coasts and in some large rivers and lakes in tropical and subtropical regions.

THEY ARE GENERALLY SLOW-MOVING FISH, spending most of their time on or near the seafloor or riverbed, and can reach an impressive size—up to 23 feet/7 meters in length in the case of the largetooth sawfish—a good proportion of which is made up of the saw on the front of the head, which gives the group its name. This saw is used mainly in feeding. It is covered in sense organs that can detect nearby prey, usually small fish. When a fish swims close, the sawfish suddenly lashes sideways, stunning or impaling the fish on its saw and then eating it.

**Despite their formidable appearance, sawfish
pose no real threat to humans.**

Human activity has had a huge impact on the sawfish population. Their meat, fins, and liver oil are all consumed, and their saws are valuable trophies or curios. In the past, there were fisheries aimed directly at them—one in Lake Nicaragua in the 1970s and early 1980s caught thousands each year until the population there collapsed. Sawfish are now either protected or too rare to be the direct targets of fishing, but they are still accidentally caught in general fisheries. They are extremely vulnerable to fishing nets, as their saws easily get entangled in them. Female largetooth sawfish do not breed until they are at least ten years old and have few young, so the species is slow to recover from declines. There are now very few places where there are thought to be sawfish populations, one of which is the Colorado–San Juan River system in Nicaragua and Costa Rica.

PRISTIS PRISTIS

CLASS: Chondrichthyes

FAMILY: Pristidae

IUCN STATUS (2013): Critically endangered

POPULATION: Unknown

FOUND IN: Tropical and subtropical coastal areas in the Atlantic and Pacific Oceans and in some rivers and lakes

NICARAGUA

200
CÓRDOBA

200
CÓRDOBA

NICARAGUA

FIRST-CLASS MAIL

FIRST-CLASS MAIL

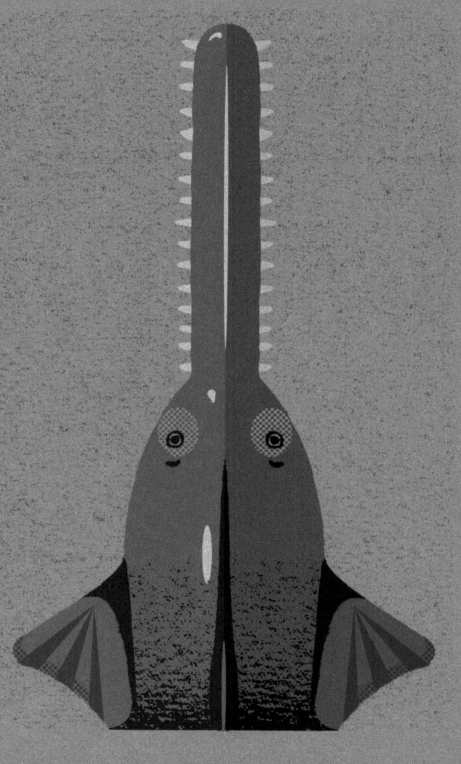

LARGETOOTH SAWFISH
PRISTIS PRISTIS

EASTERN GORILLA

The eastern gorilla, a rapidly disappearing species, is now the rarest of the great apes, apart from the Tapanuli orangutan of Sumatra. Luckily, populations in some areas are currently well protected, offering cautious reasons for optimism.

THERE ARE TWO SUBSPECIES of eastern gorillas: one found mainly in lowland areas across quite a wide area in the eastern part of the Democratic Republic of the Congo (DRC), and the other in mountains in extreme eastern DRC and adjacent parts of Rwanda and Uganda. In the range of the lowland gorilla there has been a great increase in human activity in the past thirty years, mostly due to small-scale, unregulated mining for the mineral coltan, used in cell phones and other electronic devices. This in turn has led to a massive upsurge in the numbers of gorillas being poached for meat, made worse by the numerous weapons available because of the many conflicts in the region. Surveys carried out between 2010 and 2015 indicate that the gorilla population there has shrunk by more than three-quarters in just twenty years, from around 17,000 in the early 1990s to fewer than 4,000.

The mountain gorillas are much less numerous but have fared better than the lowland gorillas.

There are two populations of mountain gorillas, separated by just 16 miles/25 kilometers of densely settled agricultural land. One is in the Bwindi Impenetrable National Park in Uganda and a small adjacent part of the Sarambwe Nature Reserve in the DRC. The other is in the Virunga volcanoes region, which spans the borders between Rwanda, Uganda, and the DRC. Intensive protection has meant that these populations have actually grown since the early 1990s, with the most recent estimates being around 400 in Bwindi and 480 in the Virungas. In both places, groups of gorillas have become accustomed to the presence of people, providing opportunities for visitors to have close encounters with these magnificent animals in their natural habitat. The resulting tourism provides income for local people and is a powerful incentive to conserve not just the gorillas themselves, but the richly diverse forests they live in.

GORILLA BERINGEI

CLASS: Mammalia

FAMILY: Hominidae

IUCN STATUS (2016): Critically endangered

POPULATION: Fewer than 5,000

FOUND IN: Democratic Republic of the Congo (DRC), Rwanda, Uganda

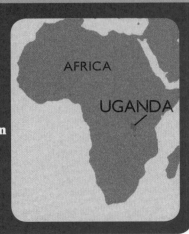

2,500 **SHILLINGS**

UGANDA

2,500 **SHILLINGS**

FIRST-CLASS MAIL

FIRST-CLASS MAIL

EASTERN GORILLA

GORILLA BERINGEI

AFRICAN WILD DOG

Like wolves, African wild dogs are social creatures. They are sometimes found in pairs but usually live in packs, nowadays of up to forty animals. In the past, when they were more common, their packs were even larger.

AFRICAN WILD DOGS ARE HUNTERS, feeding in large part on antelopes, and are found in habitats as varied as high mountain forest, desert, and even along the seashore. Mostly, though, they live in savannas and open woodlands, where there is plentiful suitable prey. Packs roam over large areas, normally from about 100 to 300 square miles/300 to 800 square kilometers, but sometimes over as much as 1,200 square miles/3,000 square kilometers. In the breeding season, when there are young pups in the dens, they keep to much smaller areas around the denning site.

Wild dogs have been widely hunted, trapped, and poisoned.

Once common in much of Africa south of the Sahara, African wild dog populations have shrunk greatly in the past century, linked directly to an expansion in human activities. People target them as predators of livestock, particularly sheep and goats, and out of fear of attacks on humans, though in reality, wild dogs pose no threat to people and can be kept away from livestock relatively easily. They are also susceptible to infectious diseases, particularly rabies and canine distemper, which can be caught from domestic or feral dogs. Although wild dog populations usually bounce back quite quickly from outbreaks of disease, isolated packs are at risk of dying out completely, with little chance of animals from other populations being able to recolonize the area.

One part of Africa that still has good numbers of the species is the vast Kavango-Zambezi Transfrontier Conservation Area, which straddles the borders of Angola, Botswana, Namibia, Zambia, and Zimbabwe, in Southern Africa. It's thought that around 130 packs, or roughly 1,300 animals in total, live there, forming what is probably the largest single remaining African wild dog population.

LYCAON PICTUS

CLASS: Mammalia

FAMILY: Canidae

IUCN STATUS (2012): Endangered

POPULATION: Estimated to be 6,000–7,000 (in 600–700 packs)

FOUND IN: Around 20 countries in Africa south of the Sahara

AFRICA

BOTSWANA

1.80 PULA

1.80 PULA

BOTSWANA

FIRST-CLASS MAIL

FIRST-CLASS MAIL

AFRICAN WILD DOG
LYCAON PICTUS

SUNDA PANGOLIN

Anyone coming across a pangolin for the first time would be forgiven for wondering what exactly they were looking at. Some kind of scaly reptile? A giant animated pinecone? Or even an artichoke?

IN FACT, PANGOLINS are highly specialized mammals that feed exclusively on ants and termites. They have no teeth but long, sticky tongues and powerful claws. Their most distinctive feature is undoubtedly their covering of large, overlapping scales of keratin, the substance that also makes up claws, hooves, and fingernails. The Sunda pangolin is one of four species of pangolin found in Asia (four others occur in Africa). It lives in tropical rain forests in Southeast Asia in a wide area from Myanmar and Thailand in the north to Indonesia in the south. Like many other species, it is under increasing threat of extinction for one main reason: hunting—in this case for its scales, which are a highly sought-after ingredient in traditional Chinese medicine.

Trade in pangolin scales has gone on for centuries.

Until recently, this trade was mostly in Asian species, but so much hunting has taken place in the past few decades that populations have collapsed virtually everywhere, and attention has now turned to African pangolins. Almost all international trade in pangolins has been illegal for many years, but this has had little effect.

The only places in Asia where pangolins remain at all common are where there is little or no hunting. One of these is Pulau Tekong, the largest of the outlying islands of Singapore. Once mostly given over to rubber plantations, and later to vegetable, fruit, and poultry farms, it is now used as a military training base. Much of it has reverted to forest, which supports a healthy population of Sunda pangolins—ample evidence that where they are not persecuted, pangolins can survive even in densely populated parts of the world.

MANIS JAVANICA

CLASS: Mammalia

FAMILY: Manidae

IUCN STATUS (2013): Critically endangered

POPULATION: Unknown

FOUND IN: Brunei, Cambodia, Indonesia, Laos, Malaysia, Myanmar, Singapore, Thailand, Vietnam

ASIA

SINGAPORE

30 CENTS **30 CENTS**

SINGAPORE

FIRST-CLASS MAIL FIRST-CLASS MAIL

SUNDA PANGOLIN

MANIS JAVANICA

MEDITERRANEAN MONK SEAL

Almost all of the world's eighteen kinds of seals live in cool or cold waters at high latitudes. The Mediterranean monk seal is an exception, its natural habitat being the relatively warm waters of the Mediterranean and Black Seas and adjacent parts of the Atlantic.

PERSECUTION BY FISHERS who accuse the seal of damaging their fishing gear, disturbances from coastal development, and hunting for its skin, oil, and meat (stretching back to Roman times) have all drastically reduced this once common animal's population and range. The Mediterranean monk seals that remain live in three widely separated areas: the Madeira archipelago, in the Atlantic; Cabo Blanco, on the coast of West Africa; and the eastern Mediterranean Sea.

There are now thought to be fewer than 1,000 Mediterranean monk seals.

The seals in the Mediterranean generally live in small, scattered groups made up of a handful of individuals. Recently, though, one substantial colony was discovered on the Greek island of Gyaros, in the Aegean Sea. This now-uninhabited island was used as a detention center for political prisoners until 1974 and then as a target site by the Greek navy until 2000. Following rumors of a population there in the early 2000s, conservationists visited in 2004 and discovered not just good numbers of seals but several pups, too, indicating that the colony was breeding healthily, unlike many other seal groups in the region. Subsequent visits have confirmed the impression of a thriving colony. The island has become part of Natura 2000, a network of nature protection sites in Europe, affording it some legal protection, and all fishing has been banned within 3 nautical miles (3½ miles/5.6 kilometers).

MONACHUS MONACHUS

CLASS: Mammalia

FAMILY: Phocidae

IUCN STATUS (2015): Endangered

POPULATION: 600–700

FOUND IN: Croatia, Cyprus, Greece, Mauritania, Madeira (Portugal), Turkey, Western Sahara

EUROPE

GREECE

AFRICA

90 CENTS

90 CENTS

GREECE

FIRST-CLASS MAIL

FIRST-CLASS MAIL

MEDITERRANEAN MONK SEAL

MONACHUS MONACHUS

KOREAN CLUBTAIL DRAGONFLY

There are around 3,000 species of dragonflies alive today, members of an ancient lineage whose ancestry can be traced back over 300 million years to the Carboniferous Period. Some species are still widespread and common, while others have limited ranges and exist in small, often declining numbers.

THE KOREAN CLUBTAIL DRAGONFLY is one of these. Once seen in several localities in the central Korean Peninsula, it has undergone a drastic decline and has been recorded recently at only one site on the Sami River in Gyeonggi province, near the demilitarized zone between South and North Korea.

Like other dragonflies, the Korean clubtail spends two or more years living underwater as a larva before metamorphosing into a winged adult. The larva's preferred habitat is slow-running stretches of lowland rivers with a sandy or muddy bottom into which it can burrow to lie in wait for prey. When the larva is ready to turn into an adult, it finds a shallow, gently sloping site and constructs a mound from which to emerge from the water. The adult then spends its short life in riverside forest.

In 2014, the dragonfly was recorded at only one site.

Unfortunately, almost all the rivers where the Korean clubtail dragonfly used to be found have been subject to intensive management in recent years. This has included dredging and bank straightening, which speeds up the water flow and destroys the larvae's silty habitat. Banks have also been made steeper, so that there are virtually no sites where the newly metamorphosed larvae can safely emerge. Part of the Sami River flows through the Korean Demilitarized Zone, where it is not subject to management, but it is not known if the dragonfly survives here in any number, as no surveys have taken place. While the rivers in the zone are relatively untouched, the forests on which the adult dragonflies depend are regularly burned to keep sight lines clear for the military on both sides.

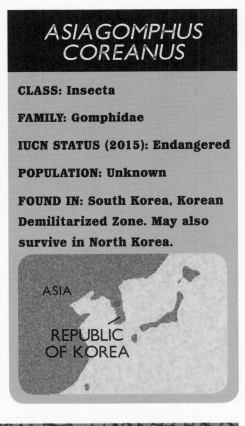

ASIAGOMPHUS COREANUS

CLASS: Insecta

FAMILY: Gomphidae

IUCN STATUS (2015): Endangered

POPULATION: Unknown

FOUND IN: South Korea, Korean Demilitarized Zone. May also survive in North Korea.

ASIA

REPUBLIC OF KOREA

REPUBLIC OF KOREA

FIRST-CLASS MAIL

FIRST-CLASS MAIL

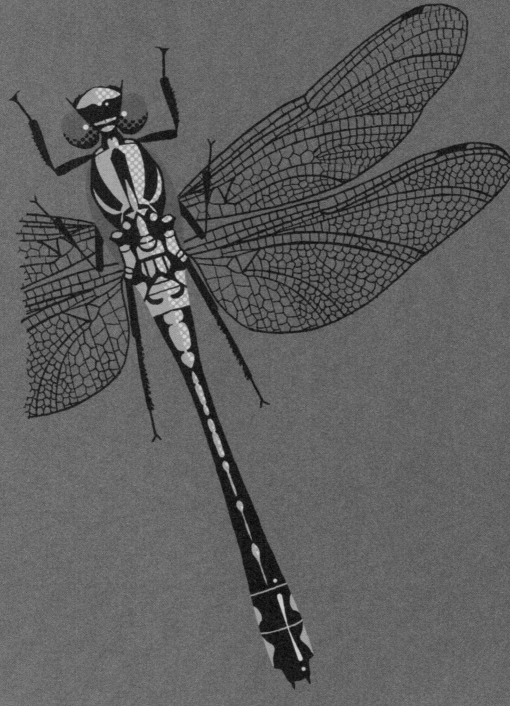

KOREAN CLUBTAIL DRAGONFLY
ASIAGOMPHUS COREANUS

KAKAPO

Before humans first arrived on New Zealand around 800 years ago, there were no land mammals apart from bats — but there were some truly remarkable birds.

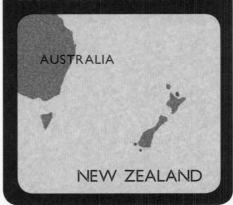

STRIGOPS HABROPTILA

CLASS: Aves

FAMILY: Strigopidae

BIRDLIFE/IUCN STATUS (2016): Critically endangered

POPULATION: 153

FOUND IN: New Zealand

AUSTRALIA

NEW ZEALAND

$1 NEW ZEALAND $1
FIRST-CLASS MAIL FIRST-CLASS MAIL
KAKAPO
STRIGOPS HABROPTILA
JAN 22 6 00 PM

SOME OF THESE BIRDS, INCLUDING the giant flightless moas, were quickly hunted to extinction by the early Polynesian settlers, but others survived, among them the kakapo, the world's largest parrot. Kakapos were quite common until the nineteenth century, when Europeans arrived and began clearing the forests and woodlands in which the kakapos lived. With even greater consequences, people also imported cats, stoats, ferrets, and weasels in a misguided attempt to control the rabbits and hares that had been introduced by humans. Unable to fly, kakapos proved defenseless against the new predators, and by the 1970s it was feared that none of the birds survived. In 1977, a handful of males were discovered in the far south of New Zealand's South Island and a bigger population on nearby Stewart Island, but there were feral cats there, and numbers were dropping fast.

By the early 1970s, the kakapo was thought possibly extinct.

In a last-ditch effort to save the species, conservationists captured and released surviving kakapos on islands where there were no cats or other carnivores. These birds started to breed, but few chicks survived because of attacks by Pacific rats. By 1995, there were only fifty-one kakapos. Protection efforts were redoubled and a host of actions were undertaken. Rats were eliminated from two of the three islands where kakapos now live, and nests on the third were rigged to scare rats away. Kakapos were given extra food to encourage them to breed, and wildlife wardens even put blankets over eggs and nestlings when their mother was off feeding to make sure they did not get too cold. By the end of 2017, the kakapo population had risen to 153. The goal now is to clear other islands of predators and introduce kakapos there too.

$1 **$1**

NEW ZEALAND

FIRST-CLASS MAIL

FIRST-CLASS MAIL

KAKAPO
STRIGOPS HABROPTILA

OKAPI

In the late nineteenth century, rumors began to circulate among scientists of a mysterious large mammal living in the rain forests of the Congo basin in Central Africa.

THE ANIMAL WAS WELL KNOWN to local people, who referred to it as an *atti* or *o'api*. Harry Johnston, governor of Uganda at the time, managed to obtain a skull of the creature and two pieces of its striped coat, which were sent to London in 1901. Based on these, the animal was scientifically named a new species of horse, but very quickly scientists realized that the okapi, as it came to be called, was not a horse but a relative of the giraffe.

Okapis remain elusive creatures to this day, rarely seen in the wild.

Considerably more is now known about okapis. They are found in only one country—the Democratic Republic of the Congo—but have a wide range within it, around 97,000 square miles/ 250,000 square kilometers. Much of this territory is inaccessible or dangerous to visit, so it is impossible to know how many okapis live there.

Okapis are unusual among large mammals in being confined to mature rain forest, where they feed on shrubs and vines in clearings where trees have fallen. They can cope with a certain amount of human disturbance but not too much, and they are absent in areas where there is extensive logging or clearing of the forest for farming. In some places okapis are also heavily hunted for their meat and pelts.

All of the signs are that okapi numbers have declined sharply, particularly in the past two decades. This is the case even in areas such as the Okapi Wildlife Reserve, which was established specifically for their protection but where forest clearance and illegal hunting continue.

OKAPIA JOHNSTONI

CLASS: Mammalia

FAMILY: Giraffidae

IUCN STATUS (2015): Endangered

POPULATION: Unknown

FOUND IN: Democratic Republic of the Congo (DRC)

AFRICA

DRC

800 Franc

800 Franc

DEMOCRATIC REPUBLIC OF THE
CONGO

FIRST-CLASS MAIL

FIRST-CLASS MAIL

OKAPI
OKAPIA JOHNSTONI

To give an idea of how endangered each of the animals in this book is, we've included the status categories used by two international conservation organizations: BirdLife and the International Union for Conservation of Nature (IUCN). These categories are determined by experts on the species in question. There are three categories for species considered to be threatened with extinction: critically endangered (the most threatened), endangered, and vulnerable (the least threatened). There are also categories for species that are extinct, extinct in the wild, or not thought to be threatened at the moment.

For each of the species, we've also included the date of the most recent expert assessment. Sometimes this was quite a while ago, and it's possible the species would be put in a different category if it were reassessed today, either because its situation has improved or declined or because better information has become available.

You can find more information on the categories and on the status of thousands of other animal and plant species at www.iucnredlist.org. For birds, you can also find information at www.birdlife.org.

To my family
M. J.

For Teresa, Harry, and Poppy
T. F.

Text copyright © 2018 by Martin Jenkins
Illustrations copyright © 2018 by Tom Frost

First U.S. edition 2019

Library of Congress Catalog Card Number pending
ISBN 978-1-5362-0543-5

19 20 21 22 23 24 TWP 10 9 8 7 6 5 4 3 2 1

Printed in Johor Bahru, Malaysia

This book was typeset in Trade Gothic, Forbes LT, and WB Endangered.
The illustrations were created digitally.

Candlewick Studio
an imprint of
Candlewick Press
99 Dover Street
Somerville, Massachusetts 02144

www.candlewickstudio.com

CANDLEWICK STUDIO
an imprint of Candlewick Press